GEOGRAPHY ALPHABET

written and illustrated by

R. M. Smith

CLARENCE-HENRY
BOOKS

Geography Alphabet
by R. M. Smith

Clarence-Henry Books • Alexandria, VA

Design and Layout by R. M. Smith

Summary: An alphabet introduction to Earth's
geographical features for kids.

ISBN-10: 0988290952
ISBN-13: 978-0988290952

First Edition
10 9 8 7 6 5 4 3 2 1

Aa is for Antarctica

Emperor
Penguins

Antarctica is the **coldest, windiest,** and **driest** place on Earth! There is **snow** and **ice** there every day of the year. The **South Pole** is in Antarctica.

Bb is for Beach

Kite

Whale

Boat

Waves

Snail

A **beach** is an area where **land** meets **water**. The **ground** is mostly made up of **sand**, **pebbles**, or **coral**.

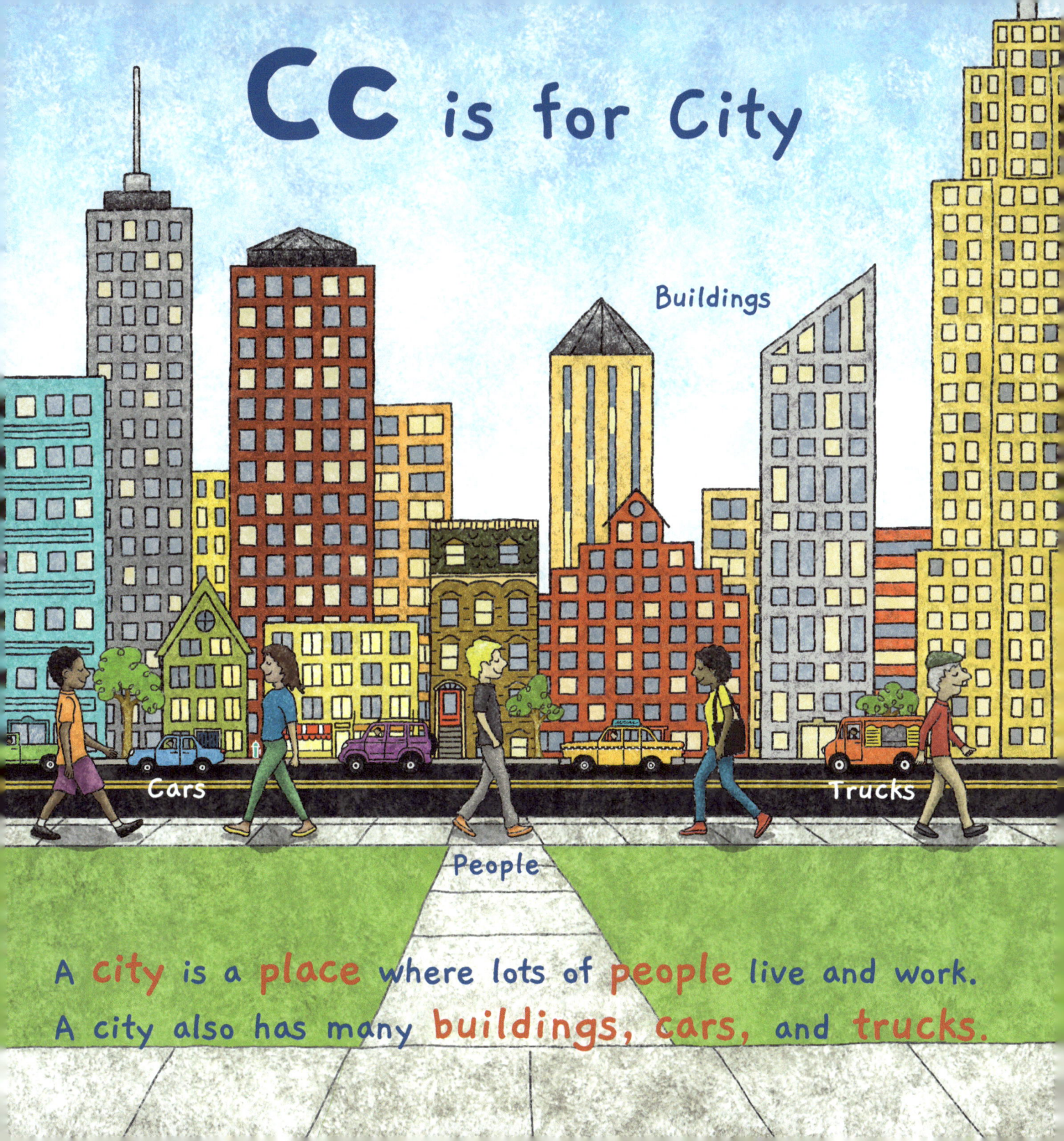

Cc is for City

Buildings

Cars

Trucks

People

A city is a place where lots of people live and work.
A city also has many buildings, cars, and trucks.

Dd is for Desert

Dunes

Cactus

Lizard

Deserts are the **driest** places on Earth. They can be very **hot** or **cold** and are made mostly of **sand** and **rock**. Most have some **plants** and **animals**.

Ee is for Equator

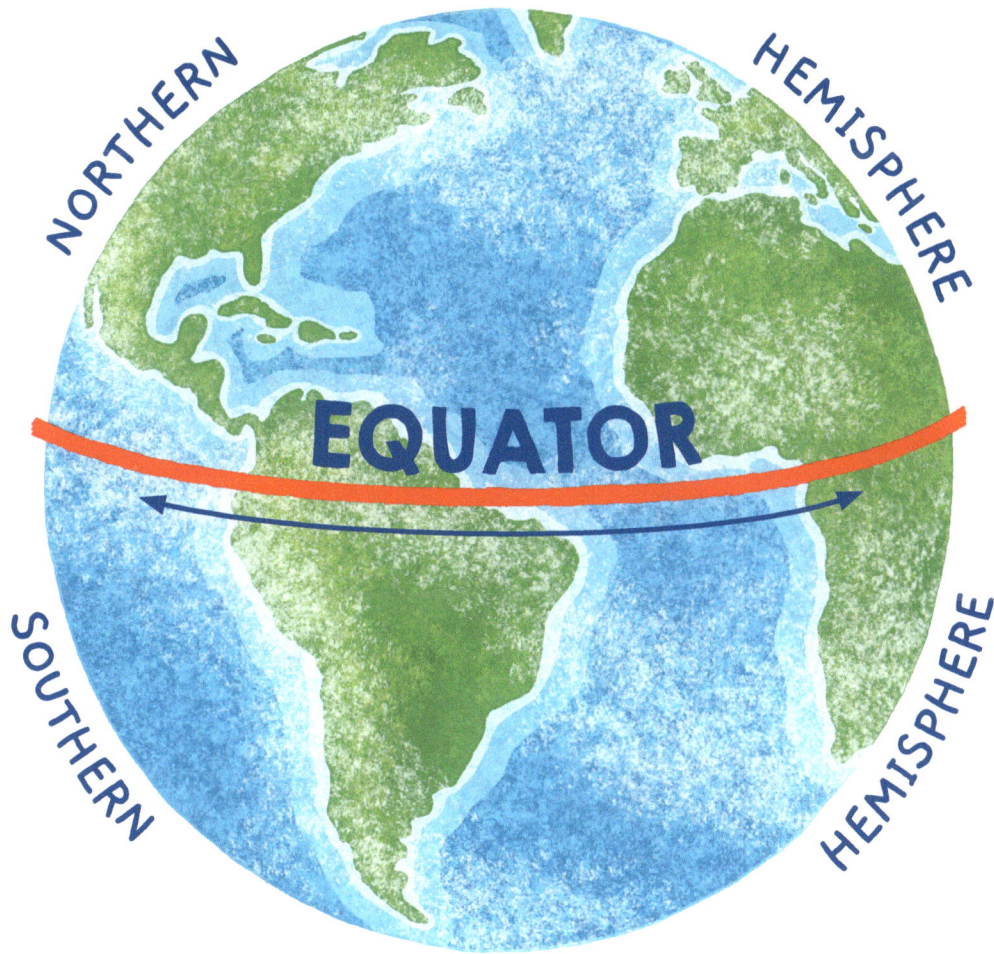

NORTHERN HEMISPHERE

SOUTHERN HEMISPHERE

EQUATOR

The **equator** is an imaginary line drawn around the **middle** of the **Earth.** The **upper half** is known as the northern hemisphere and the **lower half** is the southern hemisphere.

Ff is for Forest

Fox

A **forest** is an **area** that is filled with a **lot** of **trees**. Most **trees** in forests are really **tall**.

Gg is for Glacier

Arctic Tern

A **glacier** is a large area of thick **ice**. It **moves** really **slow**. It moves so **slow** that you can't **see** it **move**.

Hh is for Highway

Truck

Car

Cow

REST
AREA
AHEAD

GO

A **highway** is a main **road** that is made for **cars** and **trucks.**

Drivers use highways to get from one **place** to **another.**

Ii is for Island

Sea Gull

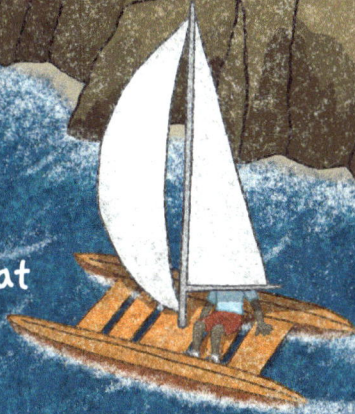

Boat

An **island** is **land** that is surrounded by **water**. It can be **large** enough for a country full of people, or **small** like the one seen here.

Jj is for Jungle

Sloth

Monkey

Toucan

Jaguar

Tree Frog

Jungles are **hot,** with lots of different **plants** and **animals.** Also, it **rains** a lot there, making the jungle a very **wet** place.

Kk is for Key

Key

◉ City

▲▲ Mountain

🌲 Parkland

– – – Hiking Trail

Lake

Road

╫╫╫ Railroad

∼∼ River

North
West ←→ East
South

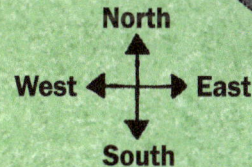

A **key** is used to help find **places** on a **map**.
It helps to spot things like cities, mountains, roads,
lakes, and more. It is also known as a **legend**.

Ll is for Lake

Coyote

A **lake** is a large body of **water** surrounded by **land**.
It's **smaller** than an ocean but **bigger** than a pond.

Mm is for Mountain

Moon

A **mountain** is a **big**, **high** piece of **land** that sits way above other land. Some mountains are so **high** that they always have **snow** on them, like this one.

Nn is for National Park

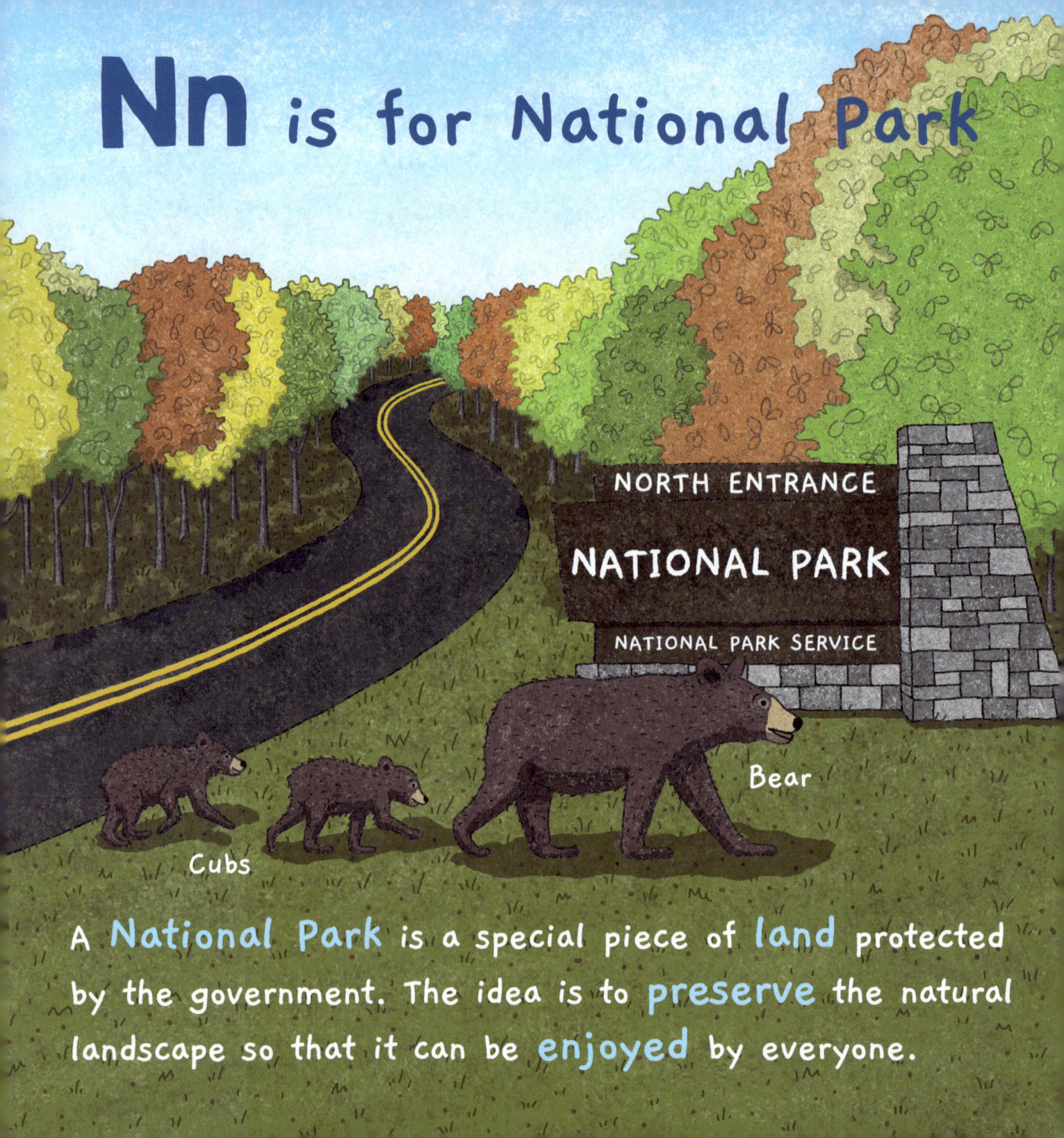

NORTH ENTRANCE

NATIONAL PARK

NATIONAL PARK SERVICE

Bear

Cubs

A National Park is a special piece of land protected by the government. The idea is to preserve the natural landscape so that it can be enjoyed by everyone.

Oo is for Ocean

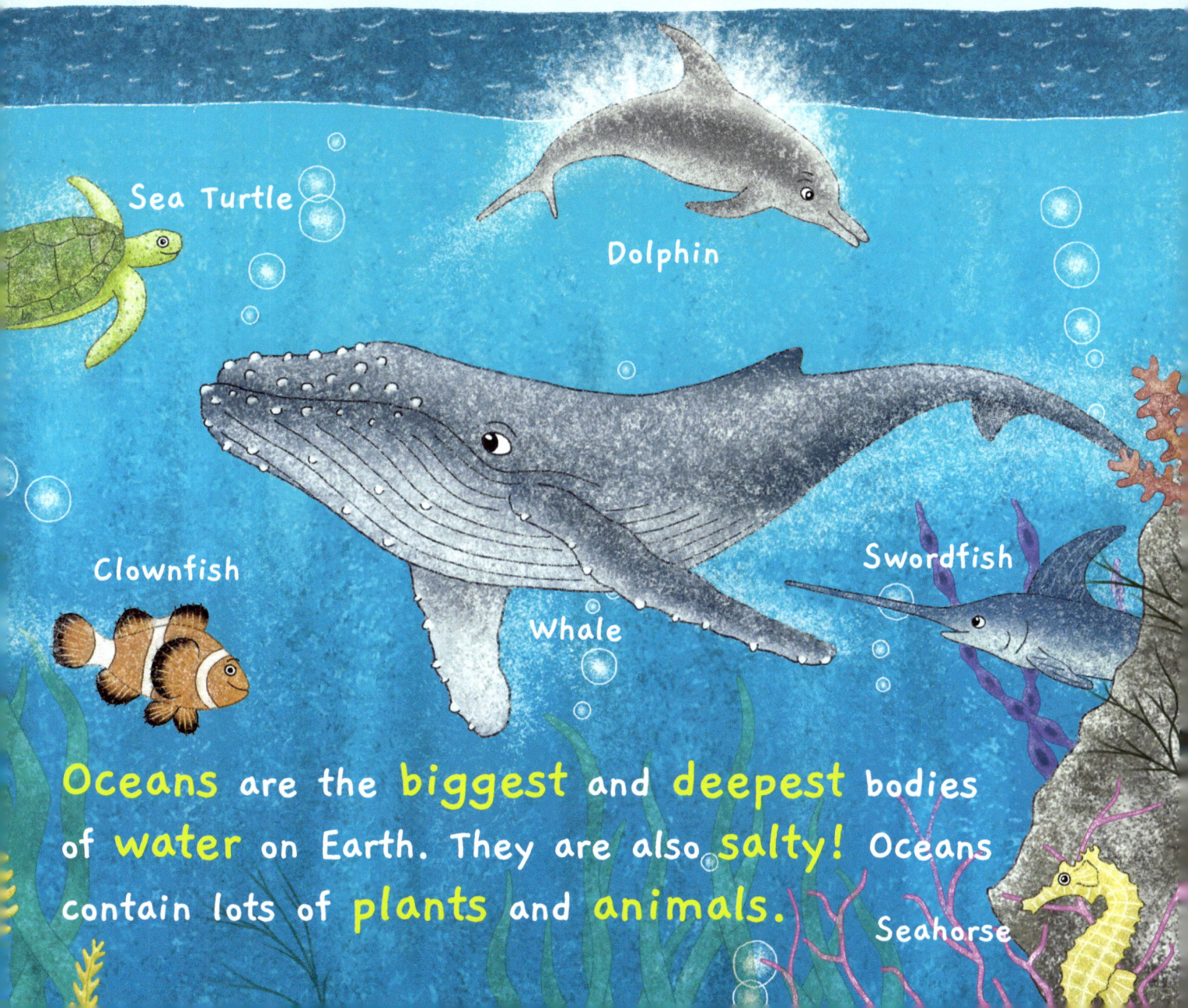

Sea Turtle

Dolphin

Clownfish

Swordfish

Whale

Seahorse

Oceans are the biggest and deepest bodies of water on Earth. They are also salty! Oceans contain lots of plants and animals.

Pp is for Plain

Windmill

Barn

Hay Bale

Buffalo

A **plain** is a large but mostly **flat** area of **land** with few trees or hills. It is great for **farming**.

Qq is for Quicksand

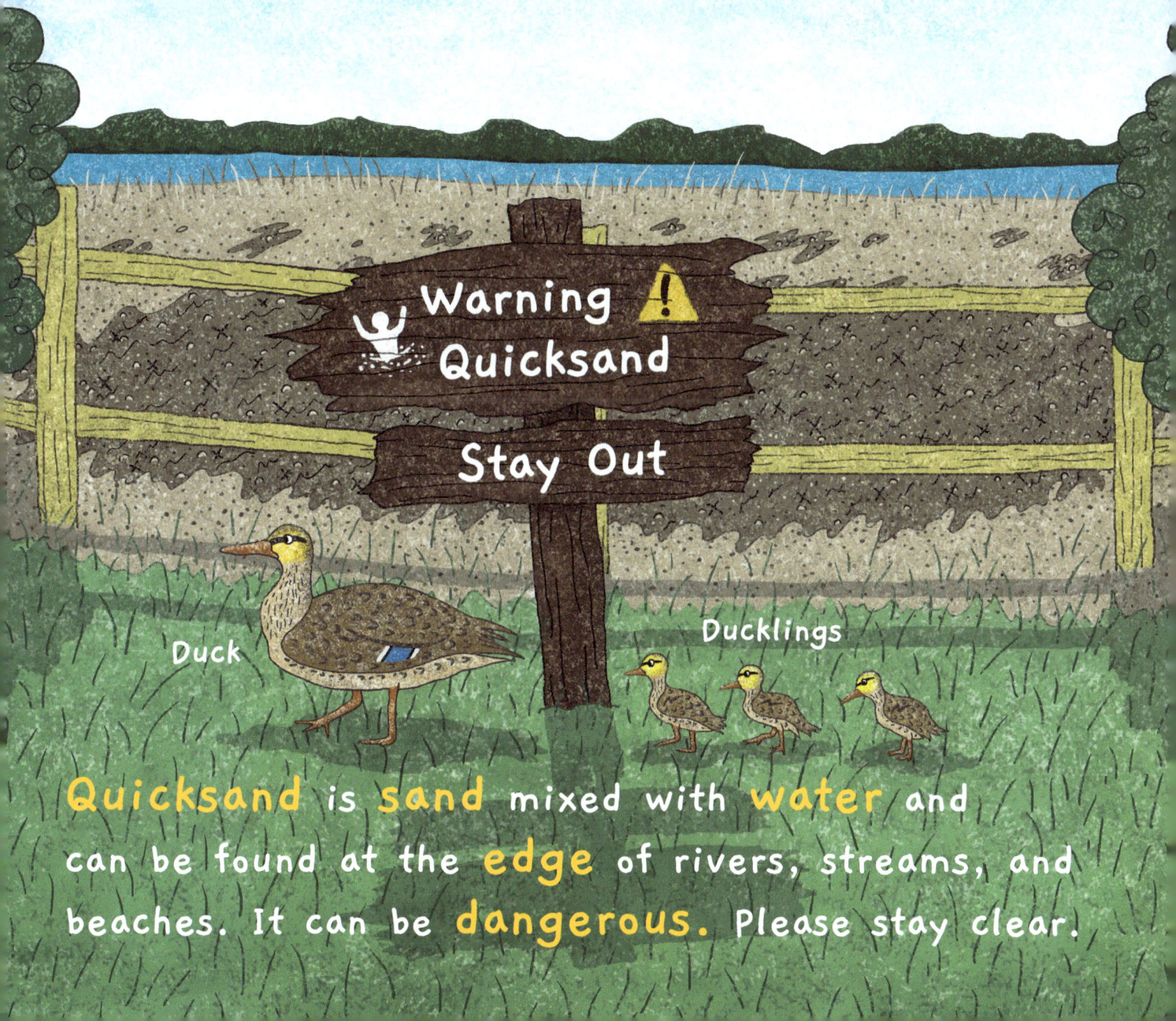

Warning ⚠️
Quicksand

Stay Out

Duck

Ducklings

Quicksand is sand mixed with water and can be found at the edge of rivers, streams, and beaches. It can be dangerous. Please stay clear.

Rr is for River

Water tower

Town

Horses

A **river** is a **long** body of flowing **water** that begins as **one** or **many** small **streams** and **grows** until it flows into seas, gulfs and oceans.

Ss is for Swamp

 Dragonfly

 Egret

Turtle

 Bullfrog

Alligator

A **swamp** is an area of **wetland** with **wild plants**, **trees**, and **animals**. Water in a swamp moves very **slowly**.

Tt is for Tundra

Reindeer

A **tundra** is a **cold**, **dry** area of land in the **Arctic**, **Antarctic**, and **mountainous** regions. It has no **trees** but some **plants** and **animals** live there.

Uu is for Upland

Hawk

An **upland** is an area where the **ground** is **higher** than the area **around** it. A group of **hills** can be an upland.

Vv is for Volcano

Smoke and ash

Lava

A **volcano** is a piece of **land** that is formed by **lava**. Lava comes out of the **top**, **runs** down the sides, and hardens to form solid **rock**.

Moose

Ww is for **Waterfall**

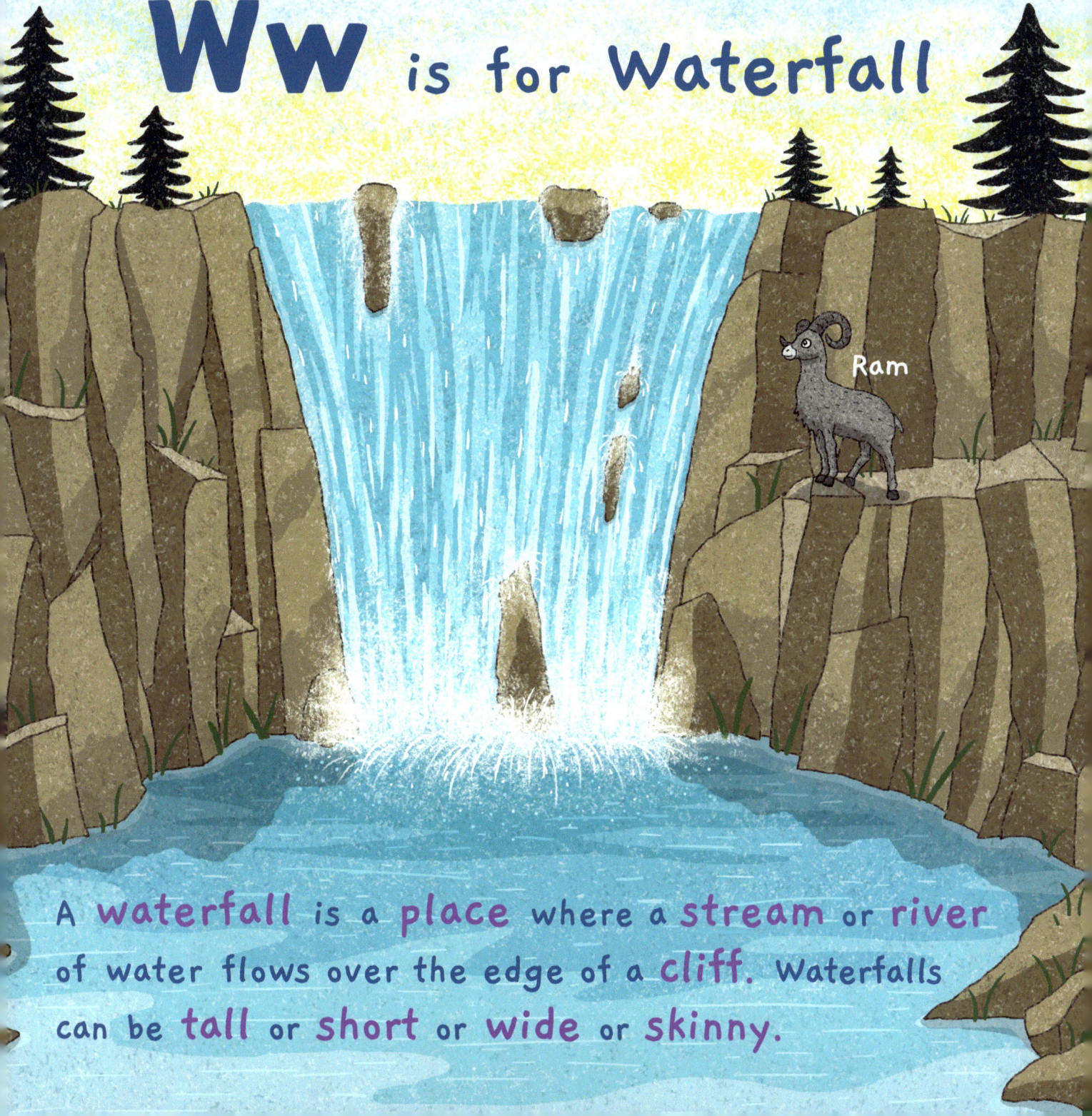

Ram

A **waterfall** is a **place** where a **stream** or **river** of water flows over the edge of a **cliff**. Waterfalls can be **tall** or **short** or **wide** or **skinny**.

Xx marks the spot

Treasure Trail

On a map, the letter X shows you are here. It can also show a place where a treasure is buried.

Yy is for Yardang

Camel

A **yardang** is land shaped by **wind**. The sandier and softer parts of **rock** are loosened and **blown** by the wind to **carve** this unusual land **shape**.

Zz is for Zone

North Polar Zone (Arctic)

Cold

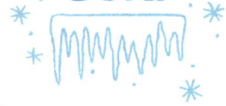

North Temperate Zone

Tropical Zone

Hot

South Temperate Zone

Cold

South Polar Zone (Antarctic)

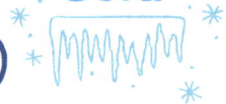

The zone shows the five main areas of the Earth by climate and geography.

www.ingramcontent.com/pod-product-compliance
Lightning Source LLC
Chambersburg PA
CBHW041241020426

42333CB00002B/41

9780988290952